EMPLOYEE'S RECEIPT

I acknowledge receipt of Keller's *Employee Food Safety Handbook*, which covers 11 different food safety topics and a glossary of food safety-related terms. The topics covered in this handbook include:

- Bacteria
- Foodborne Illness
- Personal Hygiene
- Hazard Analysis & Critical Control Points (HACCP)
- Sanitation
- Time & Temperature Controls
- Foreign Material
- Cross Contamination
- Allergens
- Pest Control
- Security
- Glossary of Food Safety-Related Terms

AF685843

____

Employee Name (Please Print)

____

Employee's Signature Date

____

Company

____

Company Supervisor's Signature

> NOTE: This receipt shall be read and signed by the employee. A responsible company supervisor shall countersign the receipt and place in the employee's training file.

Employee Food Safety Handbook,

3rd Edition

J. J. Keller & Associates, Inc.
3003 W. Breezewood Lane, P.O. Box 368
Neenah, Wisconsin 54957-0368
Phone: (920) 722-2848
www.jjkeller.com

Library of Congress Catalog Card Number: 98-85795

ISBN 1-59042-285-6

Canadian Goods and Services Tax (GST) Number: R123-317687

Printed in the U.S.A.

3rd Edition, Second Printing, September 2005

Due to the constantly changing nature of government regulations, it is impossible to guarantee absolute accuracy of the material contained herein. The Publishers and Editors, therefore, cannot assume any responsibility for omissions, errors, misprinting, or ambiguity contained within this publication and shall not be held liable in any degree for any loss or injury caused by such omission, error, misprinting or ambiguity presented in this publication.

This publication is designed to provide reasonably accurate and authoritative information in regard to the subject matter covered. It is sold with the understanding that the publisher is not engaged in rendering legal, accounting, or other professional service. If legal advice or other expert assistance is required, the services of a competent professional person should be sought.

The Editorial Staff is available to provide information generally associated with this publication to a normal and reasonable extent, and at the option of and as a courtesy of the Publisher.

TABLE OF CONTENTS

INTRODUCTION TO THE EMPLOYEE

Americans expect many things from their food supply. They want variety, quality, nutritional value, and safe food at a reasonable cost. To many consumers, safe food means that there will be no danger from hazards that could cause them illness or injury.

Therefore, preventing the contamination of food is the heart of food safety laws that regulate our food industry, and your company must continually comply with all food safety laws mandated by the federal government.

That's why your job is very important. As a person employed in the food industry, you help implement these laws at your company to make food for millions of people to eat. Because your job touches so many lives, you are responsible for providing food to consumers that is safe to eat.

Even though we're constantly aware of the importance of food safety, sometimes we develop poor habits or behaviors that could have potentially serious consequences for consumers. These consequences could be illness and even death. That's why you need to commit yourself to following all food safety rules, supporting new food safety programs, and identifying opportunities to improve food safety at your company.

This handbook provides valuable food safety information that will help you to keep yourself, the food you work with, and those who eat it safe.

BACTERIA

As an employee who handles food on a daily basis, you need to understand a little about bacteria. Bacteria are living things too small to be seen with your naked eye. They must be viewed with the help of a microscope, and therefore; are also referred to as *micro*organisms.

As a result, our daily encounters with them usually go unnoticed, but human and bacterial life interact on many levels. This certainly is true at a food company. The relationship between bacteria and food can be good or bad depending on your point of view.

Some bacteria can actually be beneficial to humans. For example, bread, wine, cheese, yogurt, beer, pickles, and summer sausage are all infused with bacteria to give a desired aroma, flavor, or texture to the food. Unfortunately, other bacteria are responsible for the spoilage and decay of food and serious foodborne illnesses.

Harmful bacteria

Bacteria, or pathogens, that cause foodborne illnesses have long, strange names, but they are not complicated to understand. Pathogens either infect people directly when ingested, or produce toxins in the packaged product or in a person's intestines after contaminated food is eaten. The results can be deadly!

Many foodborne illnesses are caused by these harmful bacteria:

- *Clostridium botulinum,*
- *Clostridium perfringens,*
- *Escherichia coli* (*E. coli*),
- *Listeria monocytogenes,*
- *Salmonella,*
- *Staphylococcus aureus,* and
- *Campylobacter.*

The following is basic information you need to know about each these potentially dangerous bacteria.

Clostridium botulinum

Clostridium botulinum is typically found in soil, water, vegetables, grains, and the intestines of animals. Under the right conditions, this bacteria will grow and produce a toxin that is very poisonous. The disease botulism occurs when the toxin is eaten.

Botulism typically causes:

- Blurred or double vision,
- Dry mouth,
- Difficulty swallowing, and
- Paralysis of respiratory muscles.

Botulinum spores will produce toxin when growing conditions are optimal. For example, spores present on vegetables at the time of harvest can be difficult to remove completely by washing. When the contaminated vegetables are bottled or canned and steamed in a cooker that does not reach the necessary pressure and temperature, some *botulinum* spores may survive because they are highly resistant to heat. All the air once present in the containers is usually removed during the cooking process. Since *botulinum* spores only grow when there is no oxygen in a container, they can quickly produce the toxin when the containers are stored at room temperature.

Therefore, botulism poisoning occurs most often with improperly canned low-acid foods or with under-processed home-prepared foods.

Botulism poisoning can be prevented through proper methods of preserving and handling bottled or canned food. Also, maintaining a low pH and using the correct time and temperature parameters during processing are essential to control this extremely dangerous pathogen.

Clostridium perfringens

This bacteria, like *Clostridium botulinum*, also produces spores that grow only in little or no oxygen. It is found in soil and in the intestines of healthy people and certain animals. It is difficult to destroy this bacteria because its spores can grow very quickly.

Clostridium perfringens poisoning can occur when inadequately cooked meats and fish are eaten. This is often the result of food being left for long periods of time in steam tables, buffets, or at room temperature. Symptoms of this illness include nausea and severe diarrhea.

It's especially important that food be kept at proper temperatures at all times. Keep hot food hot, and keep cold food cold. As a result, bacteria don't have a chance to grow.

Escherichia coliform (E. coli)

E. coli is one of the most widely known and common bacteria. It was first recognized as a cause of illness in 1982 during an outbreak of severe bloody diarrhea that was traced to contaminated hamburgers. Since then, *E. coli* has become a serious public health concern.

E. coli causes diarrhea or bloody diarrhea, stomach cramps, fever, and nausea. Usually, there is a brief time of discomfort or pain for infected people, but occasionally the diarrhea induces dehydration severe enough to cause even death. Small children and the elderly are most susceptible to infection because of their typically weak immune systems.

You can get *E. coli* infection by eating contaminated food, uncooked fruits and vegetables, unpasteurized milk and juice, and raw or rare ground beef. Typically, transmission usually occurs by cross contamination of feces to food. This can be from animal feces at the slaughterhouse or on the dairy farm. But food can also be contaminated with *E. coli* from human feces that are not thoroughly washed off your hands after using the toilet.

All good personal hygiene practices, like handwashing, must be strictly followed to prevent *E. coli* infections.

Listeria monocytogenes

Listeriosis, a serious infection caused by eating food contaminated with the bacteria *Listeria monocytogenes*, is recognized as a public health problem in the United States. An estimated 2,500 persons become seriously ill with this foodborne illness each year in this country. Of those people that become infected, 500 die.

Listeria is an extremely hardy bacteria because it can survive in various conditions. It is resistant to cold, heat, salt, pH extremes, and even stomach bile. *Listeria* is found nearly everywhere—in soil, water, and the intestines of healthy people and animals.

Contamination of food with these bacteria can happen in various ways. Vegetables can become contaminated with *Listeria* from the soil or manure used as fertilizer. Uncooked meat and dairy products can be contaminated with the bacteria when the animal is slaughtered or when milk is collected. *Listeria* can also be found in processed food, such as soft cheeses, hot dogs, and deli meats, which have been contaminated after processing.

Listeriosis has flu-like symptoms of fever, muscle aches, chills, and sometimes nausea. If the infection is severe, an intense headache, stiff neck, vomiting, delirium, and coma can occur.

Because of the severity of listeriosis, it is extremely important to control these bacteria through proper food safety controls. *Listeria* is killed by cooking, pasteurization, or other heat treatments. However, unless handwashing and sanitation procedures are followed, you can contaminate product.

Salmonella

Most of us have heard of *Salmonella*. This bacteria can be found in raw eggs, milk, meat, poultry, dairy products, and seafood.

When food contaminated with *Salmonella* is eaten, it causes *salmonellosis*. Like other foodborne illnesses, this disease can cause:

- Headaches
- Stomach pain
- Diarrhea
- Nausea
- Vomiting
- Fever

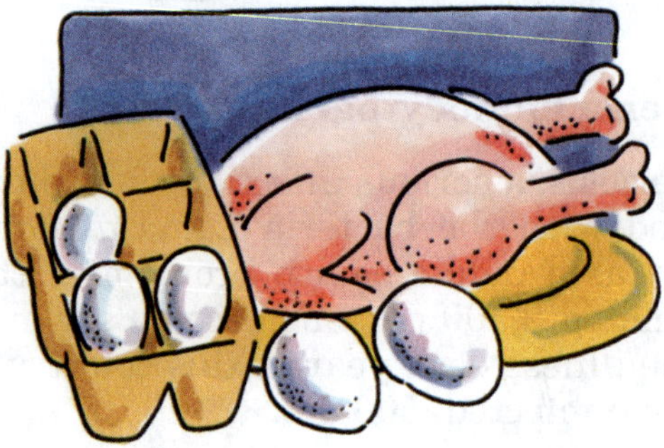

Correct processing controls, like sufficient heat treatment or the rapid development of acid in a fermentation process, are important to prevent *salmonellosis.* Avoiding cross contamination from raw to cooked food and maintaining correct storage temperatures are adequate food safety measures as well.

Staphylococcus aureus (Staph)

Staph is not found in the same places as the other types of bacteria already mentioned. Rather, it's usually found in our noses, throats, infected cuts, and on our skin. Therefore, *Staph* can be transferred to food through poor personal hygiene habits.

Staph produces a toxin that is heat resistant, so it may not be destroyed during a heat process, like cooking. Therefore, cooked food may still cause illness from this bacteria.

Poor refrigeration or inadequate heating methods speed up the growth of *Staph*. That's why it's very important to store ingredients at proper temperatures and to follow specific time and temperature controls while cooking product.

In addition, you must practice good personal hygiene to prevent this bacteria from spreading. Wash your hands with sanitizing soaps or solutions provided at handwashing stations. Do not touch your face, hair, earplugs, etc. while working directly with product.

Thoroughly cover any cuts, sores, or boils and let your supervisor know about them.

Campylobacter

When it comes to foodborne illness, big outbreaks make headlines. Like *E. coli* in apple juice and alfalfa sprouts, *Listeria* in cheese and hot dogs, and *Salmonella* in eggs and on poultry. But the most frequently diagnosed foodborne bacteria rarely makes the news. What's the name of this unsung bug? *Campylobacter.*

Most *Campylobacter* infections are sporadic and not associated with an outbreak, but it causes up to 4 million infections a year. The bacteria is commonly found in the intestinal tracts of people or animals without causing any symptoms of illness. But eating contaminated or undercooked poultry or meat, or drinking raw milk or contaminated water, may cause *Campylobacter* infection, or campylobacteriosis.

Symptoms of campylobacteriosis usually occur within two to 10 days of ingesting the bacteria and include mild to severe diarrhea, fever, nausea, vomiting, and abdominal pain.

Most people infected with *Campylobacter* can get well on their own without treatment, though antibiotics may be prescribed for severe cases. But complications can occur, such as urinary tract infections or meningitis. The bacteria is now recognized as a major contributing factor to Guillain-Barré syndrome, the most common cause of acute paralysis in both children and adults.

Spoilage bacteria

Pathogens that can cause dangerous foodborne illnesses are just one type of bacteria. There are also bacteria that cause food to deteriorate and develop unpleasant odors, tastes, and textures. These are called spoilage bacteria and can cause fruits and vegetables to get mushy or slimy, or meat to develop a bad odor.

Most people would not choose to eat spoiled food. However, if they did, they probably would not get sick. Remember, harmful bacteria like *E. coli*, *Salmonella,* and *Listeria* cause illness, but generally do not affect the taste, smell, or appearance of food. Food that is left too long at unsafe temperatures could be dangerous to eat, but smell and look just fine.

There are different spoilage bacteria and each grows at specific temperatures. Some can grow at the low temperatures under refrigeration or freezing. Others grow well at room temperature. Bacteria will grow anywhere they have access to nutrients and water.

Under the correct conditions, spoilage bacteria reproduce rapidly and the populations can grow very large. In some cases, they can double their numbers in as little as 30 minutes. The large number of these bacteria and their waste products cause the objectionable changes in odor, taste, and texture.

A food company is primarily concerned about controlling the growth of harmful bacteria. But, spoiled food is very unappealing to customers. Therefore, it's important that you continue to guard against all bacteria contamination to help make a safe and quality product.

Prevention

Ways to prevent the contamination of and control bacteria in food include:

Time and temperature controls — Time and temperature controls prevent or slow down the growth of bacteria.

Moisture — If a product is dry or has little water, then bacteria can't grow as easily.

Sanitation — If equipment is kept clean of food residue, it takes away bacteria's food source, preventing them from growing.

Acidity — Bacteria do not like very acid conditions and, therefore; food is often preserved by adding acid (e.g., pickling).

In addition, your company may test batches of food and the surfaces of equipment in an effort to detect bacterial contamination.

In conclusion

Bacteria are naturally present everywhere in our environment. This includes your workplace. You need to follow all food safety rules so bacteria doesn't contaminate product and make someone sick.

Name ____

Date ____

BACTERIA REVIEW

1. Bacteria are:
   a. Living organisms so small they can only be seen with the help of a microscope
   b. Living organisms large enough to be seen with the naked eye
   c. Organisms that are made of "micros"
   d. None of the above

2. Bacteria are found in:
   a. Food
   b. Soil
   c. Intestines of people and animals
   d. All of the above

3. Bacteria are used to make:
   a. Bread
   b. Cheese
   c. Summer sausage
   d. All of the above

4. Which is not a harmful bacteria:
   a. Listeria
   b. E. coli
   c. Zoology
   d. Salmonella

5. Disease caused by harmful bacteria is called:
   a. Microbial poisoning
   b. Foodborne illness
   c. Food contamination
   d. Food infection

6. A common symptom of many foodborne illnesses include:
   a. Vomiting
   b. Blindness
   c. Sleepiness
   d. None of the above

7. A procedure food companies use to try to control bacteria is:
   a. Refrigeration
   b. Cleaning and sanitizing
   c. Cooking
   d. All of the above

8. The most important personal hygiene practice to control the spread of bacteria is:
   a. Wearing hair restraints
   b. Washing your hands
   c. Bathing daily
   d. None of the above

9. Spoilage bacteria can cause food to become:
   a. Mushy
   b. Slimy
   c. Smelly
   d. All of the above

10. Ways to prevent the contamination of and control bacteria in food include:
   a. Time and temperature controls
   b. Sanitation
   c. Moisture control
   d. All of the above

FOODBORNE ILLNESS

Many people do not think about food safety until a food-related illness affects them or a family member. While the United States food supply is one of the safest in the world, the Centers for Disease Control and Prevention (CDC) estimates a staggering 76 million people get sick, more than 325,000 are hospitalized, and 5,000 die each year from foodborne illnesses associated with contaminated food. Preventing foodborne illnesses and deaths remains a major public health challenge.

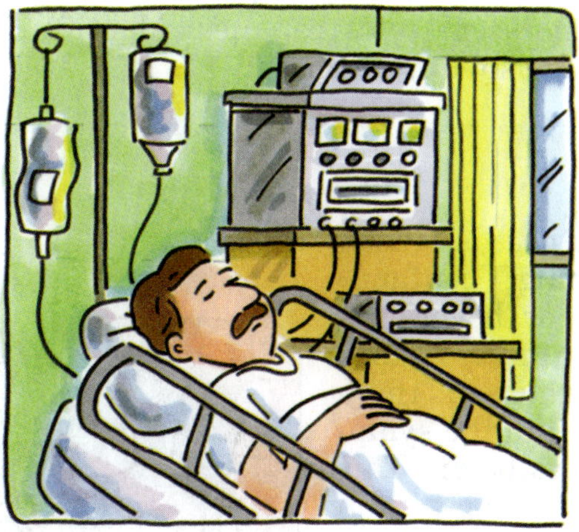

The changing world of foodborne illness

The types of foodborne illnesses are constantly changing. A century ago, typhoid fever, tuberculosis, and cholera were common. However, improvements in food safety such as pasteurization of milk, safe canning, and disinfection of water supplies have conquered most.

Today other foodborne illnesses have taken their place, including some that have only recently been discovered. In the last 15 years, several important diseases of unknown cause have turned out to be complications of foodborne infections. For example, certain strains of *E. coli* can cause kidney failure in young children and infants; *Salmonella* can lead to reactive arthritis and serious infections; *Listeria* can cause meningitis and stillbirths; and *Campylobacter* may be the most common cause for Guillain-Barre syndrome.

New microorganisms that cause disease emerge as public health problems for several reasons:

- Microorganisms can evolve;
- Our environment and the way we live are changing;
- Food production practices and consumption habits change; and
- Better laboratory tests can now identify bacteria that were previously unrecognized.

Illness

Foodborne illness is caused by consuming contaminated foods or beverages. After harmful microorganisms are swallowed, there is a delay before the symptoms of illness begin. This delay may range from hours to days, depending on the type of organism, and on how many of them were swallowed.

During this time, they pass through the stomach into the intestine, attach to the cells lining the intestinal walls, and begin to multiply there. Some stay in the intestine, some produce a toxin that is absorbed into the bloodstream, and some can invade deeper body tissues (e.g., kidneys). The symptoms produced depend greatly on the type of microorganism, but often include diarrhea, stomach cramps, and nausea.

Symptoms of foodborne illness can be mistaken for the flu. Although nausea, vomiting, and diarrhea sometimes occur during a flu infection, such symptoms are rarely prominent because flu is a respiratory illness. In fact, experts from the CDC report that many of the intestinal illnesses commonly referred to as "stomach flu" are actually caused by foodborne pathogens. People typically do not associate these illnesses with food because the onset of symptoms often occurs two or more days after the contaminated food is eaten.

Contaminated or unclean food can be very dangerous, especially to:

- Pregnant women,
- Small children,
- The elderly, and
- People with weakened immune systems (i.e., cancer patients).

Diagnosis

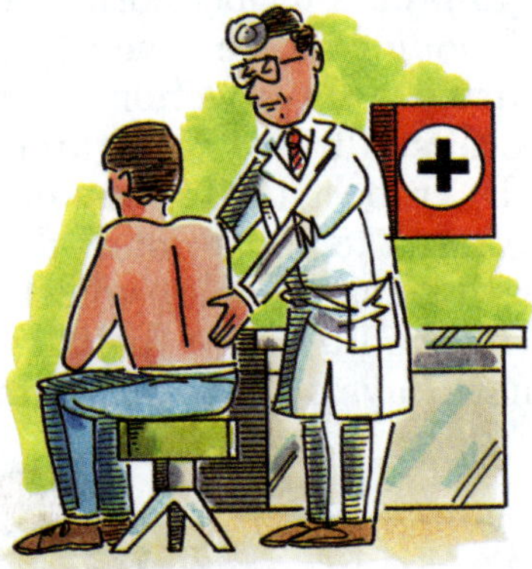

Many foodborne illnesses are not identified by routine laboratory procedures and require specialized, experimental, and/or expensive tests. If a correct diagnosis is to be made, a patient has to seek medical attention. Because many ill persons do not seek attention, many cases of foodborne illness go undiagnosed.

Treatment

There are many different kinds of foodborne diseases and they may require different treatments, depending on the symptoms they cause. The majority of cases are mild and cause symptoms for only a day or two. Illness can be treated by increasing fluid intake to replace lost fluids and electrolytes.

Unfortunately, hospitalization may be required if the illness is severe.

Prevention is the key

We live in a microbial world, and there are many opportunities for food to become contaminated as it is produced and prepared.

There is good news. Many foodborne illnesses can be prevented by various food safety practices. These include:

- Practicing good personal hygiene (e.g., handwashing),
- Properly cleaning and sanitizing equipment and work areas,
- Following time and temperature controls, and
- Maintaining a pest control program.

In conclusion

After eating contaminated food, people can develop anything from a short, mild illness, to a life-threatening disease. If food is handled and prepared safely, most of these illnesses can be avoided. That's why it's extremely important that you follow all company food safety procedures to make food that's safe to eat.

Name____
Date____

FOODBORNE ILLNESS REVIEW

1. Foodborne illness affects how many people each year:
   a. 76,000
   b. 760,000
   c. 76,000,000
   d. None of the above

2. Foodborne illness kills how many people each year:
   a. 500
   b. 5,000
   c. 50,000
   d. None of the above

3. Foodborne illnesses have been conquered due to improvements in food safety, such as:
   a. Pasteurization of milk
   b. Safe canning
   c. Disinfection of water supplies
   d. All of the above

4. New microorganisms emerge as public health problems because:
   a. Microorganisms can evolve
   b. Our environment and the way we live are changing
   c. Food production practices and consumption habits change
   d. All of the above

5. When harmful microorganisms are swallowed, they:
   a. Stay in the intestine
   b. Produce a toxin
   c. Invade deep body tissues
   d. All of the above

6. Symptoms of foodborne illness include:
   a. Diarrhea
   b. Stomach cramps
   c. Nausea
   d. All of the above

7. Contaminated food can be very dangerous, especially to:
   a. Pregnant women
   b. Small children
   c. The elderly
   d. All of the above

8. Treatment of foodborne illness typically requires you to:
   a. Eat healthy foods
   b. Get a shot
   c. Drink lots of fluids
   d. None of the above

9. Food can become contaminated by:
   a. Infected employees
   b. Cross contamination
   c. Neither a nor b
   d. Both a and b

10. Many foodborne illnesses can be prevented by:
    a. Practicing good personal hygiene
    b. Properly cleaning and sanitizing equipment and work areas
    c. Following time and temperature controls
    d. All of the above

PERSONAL HYGIENE

With all the cleaning and sanitizing we do, you would think that the risk of someone getting sick from the food you make would be extremely rare. The truth, however, is that millions of people suffer from foodborne illnesses each year. The risk of a customer eating food contaminated with bacteria is real, unless you follow good personal hygiene habits and procedures to ensure the safety of the food you help make, and the safety of the people who eat it.

If you're sick

The first rule you must always follow is never to go to work if you're sick. A fever, diarrhea, or vomiting could indicate that you have an illness that may be contagious.

Coming to work could expose your co-workers, production areas and equipment, the product, and customers to harmful infections. The best thing to do is just to stay home until you feel well enough to work again.

Personal cleanliness

Personal cleanliness begins at home. You are the only one who can maintain your personal hygiene. Several simple habits you should follow to help ensure good personal cleanliness are:

- Showering or bathing every day before you come to work. Use plenty of soap when showering in order to get really clean. Bacteria and dirt can easily be found on unclean hair and skin.
- Examining your body for cuts or sores where harmful bacteria can thrive.
- Putting on clean, laundered clothes (i.e. socks, underwear, pants, and shirt) after bathing and before going to work.
- Keeping your beard shaved or trimmed and your hair well groomed. Long hair should be in a braid, ponytail, or barrette because strands are less likely to come loose and fall into food.
- Trimming, cleaning, and filing fingernails frequently. The rough edges of cracked, chipped, or broken fingernails can contain dirt and bacteria.
- Removing nail polish, fake nails, and jewelry before starting work.

Clothing

Using the right tool for the right job is an age-old adage. Clothing isn't often thought of as a tool, but wearing the right type of clothing when you work with food can be as important as the equipment you use.

Most food industry jobs require you to wear personal protective equipment (PPE). This includes items such as hairnets, beard covers, gloves, aprons, plastic sleeves, frocks, and uniforms. PPE must always be worn in a way that protects against the contamination of food and equipment.

As an employee, you must understand and practice each of the following:

- Wear frocks or uniforms only at work. Never wear them outside the plant or at home.
- Never modify the fit or style of the garment (e.g., cutting off the sleeves) in a way that might expose your skin or other potential contaminants to product.
- Change frocks or uniforms when they become soiled or dirty, or when you go from raw to finished product areas. Dangerous bacteria can hitchhike on your clothing.
- Maintain gloves, aprons, and plastic sleeves while handling food in an intact, clean, sanitary condition at all times. If these garments become soiled or torn, they can no longer serve their intended purpose, and you must get new ones.
- Wear hairnets, beard covers, caps, and other hair restraints in a manner that appropriately covers all visible hair. Loose hair can fall into food.
- Remove and hang all smocks, coats, aprons, etc. outside of the restroom before you use it.
- Frocks and uniforms must be laundered as appropriate.

Clothing can both protect food from and be the source of food safety hazards. For example, if you wear shirts that have pockets and carry items such as cigarettes, lighters, pens, and candy, these can fall into food. When this type of contamination occurs, it could lead to serious consequences for customers.

It's important to scrutinize the type of clothing you wear for work. Try to avoid loose-fitting clothes, shirts with pockets, hooded shirts with dangling strings, and sleeves with buttons or snaps. All of these garments could be a danger to both you and product.

Footbaths

Your company may have sanitizing footbaths that you must walk through before entering a production area. Never skip the footbath just because you're in a hurry or think your shoes or boots aren't dirty. Contaminating your work area could be deadly for customers.

If possible, leave your work boots or shoes in a locker or designated storage area before leaving work. This helps keep potential contaminants outside the plant.

Handwashing

You carry millions of bacteria on your hands. Most are harmless, but some can cause serious illnesses. When you forget to wash your hands, or don't wash them properly, you can easily spread these bacteria to the product you work with and eventually make somebody very sick.

Therefore, handwashing is one of the most important functions you perform during your daily work routine. Handwashing, when done correctly, is the single most effective way to prevent the spread of disease. Proper handwashing technique is easy to learn and can significantly help reduce the contamination of food.

Handwashing seems pretty easy; after all, you've been doing it for years. But proper handwashing requires more than simply soaping your hands under running water.

You should follow these simple handwashing steps to eliminate as many bacteria from your hands as possible:

1. Turn the water on to the warmest temperature you can tolerate and wet your hands under the running water.

2. Take an adequate amount of soap or sanitizer that's provided.

3. Scrub your hands vigorously against each other for at least 20 seconds, making sure that the soapsuds cover and clean every part of your hands. Pay particular attention to the webbing between your fingers and the area under and around your fingernails where bacteria hide.

4. Rinse your hands thoroughly under the hot running water, making sure that all traces of soap are completely rinsed from your hands.

5. Dry your hands with the disposable, single-use paper towels that are provided. Throw the paper towels in the appropriate trash receptacle.

Pay special attention to what you're doing and wash your hands extremely thoroughly.

When you wash your hands is just as important as how you wash them. You must wash your hands after any activity that could contaminate the food or equipment you work with. Wash your hands at any of the following times:

- Immediately before starting at your workstation prior to putting on gloves, plastic sleeves, armguards, aprons, etc.
- After coughing or sneezing into your hands.
- After touching or scratching any part of your skin, hair, eyes, or mouth.
- After adjustments are made to your clothing, such as coats, hard hats, hairnets, earplugs, shoes, etc.
- After using a tissue or handkerchief to wipe or blow your nose.
- After eating, drinking, or smoking.
- After using the restroom for any reason.
- After picking an item up off of the floor.

- Before and after handling raw meats, poultry, or other raw foods.
- After handling items such as boxes, labels, garbage, brooms, hoses, etc.
- After each absence from your workstation.

Also, if your work area has a hand dip station, use it frequently throughout the day.

Jewelry

A very important personal hygiene rule to follow is to not wear jewelry. Remove all of it before you start work. This includes watches, earrings, necklaces, bracelets, and rings.

Jewelry can harbor skin, debris, hair, oil, and harmful bacteria. That's because jewelry rests next to your skin and hair when you wear it, collecting potential contaminants in it's grooves and crevices.

Also, jewelry can break and fall into food. An earring may come apart or the face of a watch could pop off. These pieces could harm a customer if eaten.

Either leave your jewelry at home or store it in your locker. If you're not able to remove some pieces of hand jewelry, like a wedding ring, you may have to wear a rubber glove or other covering over it. The covering must remain intact and be kept clean.

Medic alert jewelry can stay on at all times. In a food plant, the preferred type of medic alert jewelry is a chain with a medic alert pendant, worn under your shirt.

Gloves

Although handwashing removes most of the bacteria on your hands, it does not remove all of it. Therefore, gloves are commonly used to provide a barrier between the bacteria on your hands and the product you touch.

If you wear gloves when handling product, you still need to make sure that you thoroughly wash your hands before gloving. Otherwise, the contamination on your hands will simply be transferred to your gloves. **Remember, gloves are used to protect product, not your hands.**

If your job requires you to wear gloves, they must be kept in an intact, clean, and sanitary condition. When your gloves get dirty or torn, throw them away and put on a new pair. Also, if hand dip stations are provided, use them throughout the day.

Gloves are not a magic shield of protection on your hands. They can spread bacteria in the same way that your bare hands do. So change your gloves whenever you would wash your hands.

Also, never put gloves in your pockets or take them into restrooms, lunchrooms, offices, or other unclean areas. Leave them at your workstation when you go on break or leave your work area. Get a new pair of gloves as required.

Hair restraints

One of the most important pieces of clothing you wear on the job is a hair restraint. Harmful bacteria found on your skin and in your hair can easily be transferred to product. Again, these bacteria can make people sick.

It is your responsibility to wear a hair restraint effectively, making sure it covers your hair completely so none can escape and fall into food. If you have long hair, you should first put it in a ponytail or braid. This way, strands are less likely to come loose.

If you have a beard, a restraint should be worn for the same reasons as a traditional hairnet. Beard hair can just as easily contaminate product. Special beard covers are available to restrain beard hair, and must be worn along with traditional hairnets or other head coverings.

You should never do any personal grooming, like combing your hair or smoothing your beard, while working with product. Try not to touch your hair until your shift is complete or you're in a locker room, changing area, restroom, or break area.

If you have to adjust your hair restraint or touch your hair, follow these simple steps:

1. Use a restroom or changing area to make sure your hair is completely under your hair restraint,
2. Wash your hands before returning to your work area, and
3. Put on a new pair of gloves as necessary.

Personal belongings

Your personal belongings (i.e., purse, jacket, jewelry, cigarettes, magazines, newspapers, books) should be kept in a designated storage area, such as a locker room.

Lunches, snacks, and candy must also be stored in designated areas. However, these items shouldn't be stored with clothing and other personal belongings because they could attract pests. Contamination from these pests could be transferred from your clothes to product you make.

Many companies have lunchrooms or break areas where you can store these items in refrigerators or on shelves. Eat, drink, and smoke only in these designated areas as well. Make sure you are aware of which areas of the company are designated for this specific use and storage.

The proper storage and use of personal belongings is strictly enforced at food companies to prevent contamination of product and equipment.

In conclusion

Understand that maintaining good personal hygiene and following all hygiene rules at work keeps food safe for customers to eat.

Name ____

Date ____

PERSONAL HYGIENE REVIEW

1. Personal hygiene is:
   a. Keeping yourself clean
   b. Sanitizing of equipment and food work areas
   c. Coming to work on time
   d. None of the above

2. Which is not a good personal hygiene practice:
   a. Bathing daily
   b. Wearing hair restraints
   c. Wearing clean jewelry
   d. Handwashing

3. Good personal hygiene practices include:
   a. Not working when sick or infectious
   b. Wearing clean clothes to work
   c. Trimming, cleaning, and filing fingernails frequently
   d. All of the above

4. Which item can be brought into food work areas:
   a. Chewing gum
   b. Hair restraints
   c. Lunches
   d. Candy

5. Personal protective equipment (PPE) includes:
   a. Gloves
   b. Plastic sleeves
   c. Smocks
   d. All of the above

6. When wearing personal protective equipment (PPE), you should:
   a. Modify the cut or style
   b. Keep it clean and uncontaminated, and change it when it becomes torn or soiled
   c. Keep it on until the end of your shift, even if it's contaminated
   d. None of the above

7. How many steps are involved in a proper handwashing procedure:
   a. Seven
   b. Five
   c. One
   d. Two
8. You must always wash your hands after:
   a. Using the restroom for any reason
   b. Touching or scratching any part of your skin, hair, eyes, or mouth
   c. Eating, drinking, or smoking
   d. All of the above
9. Proper handwashing technique should include which of the following:
   a. Don't worry about cleaning under your fingernails or between fingers
   b. Using the coldest water possible
   c. Using the germicidal/anti-bacterial soap or sanitizer provided
   d. Drying your hands on your uniform
10. Where can lunches, drinks, candy, and chewing gum be allowed in food facilities:
    a. In processing areas
    b. Only in lunchrooms, break areas, or other designated storage areas
    c. In clothing storage or changing room areas
    d. All of the above

HAZARD ANALYSIS & CRITICAL CONTROL POINTS (HACCP)

Have you heard of **H**azard **A**nalysis and **C**ritical **C**ontrol **P**oints? HACCP (pronounced há-sip) may sound like it is complicated, but the reason it exists is fairly simple. It is a system of food safety control.

Both the Food and Drug Administration (FDA) and the United States Department of Agriculture (USDA) have made HACCP a rule that many food companies must follow.

This chapter will provide you with valuable HACCP information, so that you can help to keep your company in compliance with HACCP requirements.

History of HACCP

HACCP was first developed in the 1960's as a spin-off of the U.S. space program. The government wanted to make sure that the food eaten by astronauts would not harm or make them sick in space. They wanted food that was completely free of any bacteria, toxins, chemicals, and physical hazards.

As a result, HACCP was soon developed as a way to eliminate these types of hazards to produce the safest food possible.

Today, HACCP is a food safety system that has been adapted by the food industry all over the world. HACCP has been embraced as critically needed to ensure the safe, wholesome production of food. The same food you and your family eat.

Overview of HACCP

HACCP focuses on identifying and addressing hazards **before** they contaminate product. This must be done from the time raw materials are received to when product is shipped to consumers. Each production step in between must be examined to identify any possible biological, chemical, or physical food safety hazards that are reasonably likely to contaminate product and cause illness or injury to a consumer.

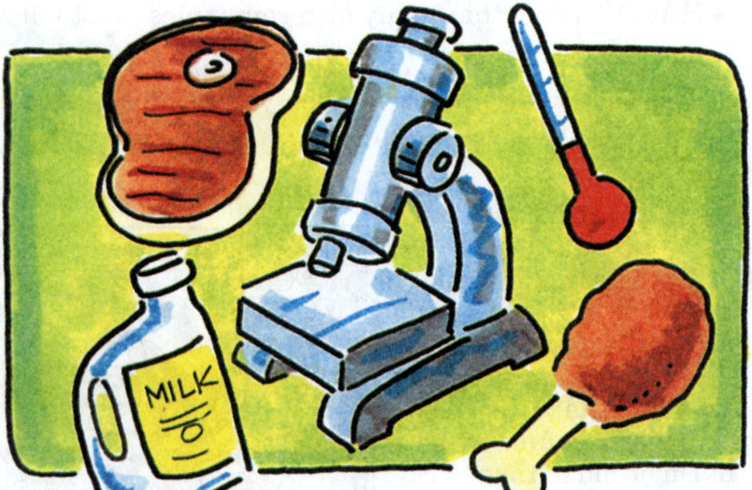

This careful analysis must be completed for every product the company makes.

Food safety measures are then put into place to eliminate or control these hazards.

The seven principles of HACCP

A HACCP plan is carefully developed using these seven principles:

1. **Conduct a hazard analysis** — the process of identifying and evaluating all of the potential biological, chemical, and physical food safety hazards associated with each step in the production process (i.e., receiving, storage, mixing, cooking, cooling, etc.) for each product made, to determine which are likely to cause illness or injury if they are not properly controlled.
2. **Identify critical control points (CCPs)** — a point, step, or procedure in the production process at which control can be applied to prevent, eliminate, or reduce a food safety hazard to an acceptable level of safety before the product leaves the company's control.
3. **Determine critical limits for each CCP** — boundaries of safety for each CCP to control the identified food safety hazard to ensure the product is safe to eat.
4. **Develop monitoring procedures** — a planned sequence of observations or measurements to make sure a CCP is under control.
5. **Establish corrective actions** — a procedure to be followed when monitoring indicates that there is a deviation from an established limit at a CCP. A corrective action tells you what to do if something goes wrong.
6. **Perform verification procedures** — activities performed to verify that the HACCP plan is adequate to control food safety hazards and that the system is operating as intended.
7. **Develop an effective recordkeeping system** — documents that the HACCP system is operating according to the written plan.

As an employee, it's possible that the work you do isn't directly associated with any of the seven principles of HACCP, but odds are that it is. Activities like taking product temperatures, checking metal detectors, and keeping certain records are all HACCP-related activities you might do for your job.

The HACCP Team

There are several preliminary tasks that need to be completed before the seven HACCP principles can be utilized.

First, a HACCP team is assembled. A HACCP team is a group of people who are responsible for developing, implementing, and maintaining the HACCP system. The team consists of people who have specific knowledge and expertise of the products and processes at the company.

A HACCP team typically includes people from a variety of areas within the company, such as engineering, production, sanitation, maintenance, and quality assurance/control. You might even participate on a HACCP team.

The HACCP team has several responsibilities. First, the team makes a list of all the products made at the company. A separate HACCP plan must be developed for each product or group of products that is made.

This information helps assist the team while they're developing the HACCP plan.

HACCP in the real world

After the HACCP team has completed the preliminary tasks discussed above, they are ready to write the HACCP plan. They apply the seven principles of HACCP to each product they've identified and described. This information is the bulk of the written HACCP plan and documentation.

Hazard Analysis

First, the team must conduct a hazard analysis to identify and evaluate the hazards for each production step that is used to make each product. A hazard is any biological, chemical, or physical agent that may cause food to be unsafe to eat.

The table below includes examples of common biological, chemical, and physical hazards.

Biological hazards	Chemical hazards	Physical hazards
Bacteria: • *E. coli* • *Salmonella* • *Listeria*	Agricultural chemicals: • Pesticides • Herbicides • Fertilizers	Metal from: • Nails • Jewelry • Thermometers
Viruses: • Hepatitus A • Hepatitis B • Norovirus	Workplace chemicals: • Cleaners/sanitizers • Paints • Lubricants	Glass from: • Bottles • Watches • Light bulbs
Parasites: • *Giardia lambia* • *Cryptosporidium* • Trichinella	Food chemicals: • Preservatives • Color additives • Flavor enhancers	Plastic from: • Box/combo bands • Packaging supplies • Broom handles
	Allergens: • Fish • Milk • Peanuts	Wood from: • Pallets • Crates • Sawdust/wood chips for smokehouses

After the HACCP team has made a list of all the potential hazards for each product, they have to decide which of these hazards must be addressed in the HACCP plan. During this stage, the severity and likely occurrence of each potential hazard is closely evaluated. Also, safety concerns must be differentiated from quality concerns and thus, for HACCP purposes, the word "hazard" refers only to food safety issues.

A thorough hazard analysis is the key to an effective HACCP system. If the hazard analysis is not done correctly, unidentified hazards may exist. Without their identification, the appropriate control measures cannot be applied, which may make food unsafe to eat.

The next step is to develop preventive measures. Preventive measures are physical, chemical, or other methods that can be used to control each hazard before they become a safety issue. For example, inspecting incoming raw materials for hazards such as wood, metal, or glass could be a preventive measure used to eliminate these hazards from further contaminating product and potentially harming a consumer. Other preventive measures used to control hazards could include time and temperature controls to kill bacteria or washing equipment between production runs to eliminate cross contamination.

The steps in the process where these preventive measures are implemented are called control points.

Critical Control Points (CCPs)

After identifying each hazard and developing preventive measures, the next step is to identify critical control points. A critical control point (CCP) is any point, step, or procedure in the process at which control can be applied to prevent, eliminate, or reduce a food safety hazard to an acceptable level of safety. The reason it is termed "critical" is because it is considered the last point in the production process in which a hazard can be addressed before the product leaves the company's control.

The following are examples of CCPs:

- Chilling food;
- Cooking to a specific time and temperature to destroy harmful bacteria;
- Product formulation controls (such as adding culture to summer sausage or rennet to cheese);

- Processing procedures (such as filling and sealing cans); and
- Metal detection.

Critical limits

After the CCPs have been identified, the HACCP team must establish critical limits for each CCP. Critical limits are considered boundaries of safety at CCPs. Therefore, if a product is outside the critical limits, it is considered potentially unsafe to eat.

Critical limits are expressed as numbers or specific parameters. For example:

- "Freezing product to an internal temperature of 40 degrees Fahrenheit."
- "Cooking product to an internal temperature of 170 degrees Fahrenheit or above."
- "Obtaining a pH less than 4.6."
- "A working metal detector that rejects a detecting wand."

Monitoring

Monitoring of critical limits at CCPs is the next principle. Monitoring is making observations or measurements to assess whether a CCP is under control. These observations or measurements must be recorded.

Monitoring is essential to a HACCP system. It warns if there is a trend towards loss of control, so the company can take action to bring a process back into control before a critical limit is violated. Remember, a product that is outside the established critical limits is considered potentially unsafe to eat.

Monitoring may be continuous (e.g., metal detection) or non-continuous (e.g., taking product temperatures).

If your job involves monitoring, always follow the established procedures you were instructed to use. Monitoring procedures must give an answer right away, so corrective action can be taken immediately if something is wrong.

If you ever suspect a problem with the results from monitoring a CCP, or are unsure how to interpret them, contact a supervisor or quality assurance/control person for help.

Corrective actions

Corrective actions are simply the procedures to be followed when failure to meet a critical limit is discovered during the monitoring process. The purpose of corrective actions is to prevent foods that may be hazardous from reaching customers.

Corrective actions must define exactly what has to be done if something goes wrong. This includes indicating the responsibilities of each person who may be involved. Employees trained in monitoring may also be trained to perform the appropriate corrective actions.

A corrective action must do two things:

- First, it must bring the production process back in control; and
- Second, it must tell a person what to do with the product.

Examples of corrective actions may include:

- Immediately adjusting a cooking process;

- Holding product for further evaluation;
- Stopping a production line until a HACCP-knowledgeable person has assessed the situation and determined appropriate action; or
- Using an alternate production process.

Regardless of what corrective action is taken, records must be kept that identify the specific deviation, what corrective action was taken, details of the incident, and actions planned to prevent a reoccurrence.

Recordkeeping

HACCP requires that effective recordkeeping procedures be established. Keeping records is essential and must be planned and carried out as carefully as any other HACCP step. A HACCP program will not work unless up-to-date, accurate, complete, and reliable records are maintained.

The HACCP recordkeeping requirement adheres to the creed, "if you didn't document it, you didn't do it." Under HACCP, a product without paperwork is considered contaminated. Without documentation, there is absolutely no way the company can prove the product is safe to eat.

Thus, recordkeeping serves several important purposes as follows:

- Serves as documentation of the company's compliance with its written HACCP plan.
- Allows the company to trace the history of an ingredient, processing operations, or finished product should problems arise.

- Helps identify trends in operations that could result in a deviation if not corrected.
- Helps identify and narrow the scope of a recall, if necessary.

Every employee responsible for HACCP documentation must fully understand the importance of recordkeeping. HACCP documents essentially verify that products are produced safely. A fraudulent or incomplete record is a HACCP system failure that could have severe consequences for the company and customers.

You may be involved in keeping various HACCP records. If so, be sure they are signed and dated as necessary. All information must be legible, recorded at the required times, and accurately reflect the monitoring activities you performed.

Even though keeping HACCP records may seem like extra work, remember that you are helping to ensure the safe production of food.

Verification

After the HACCP plan has been implemented, the HACCP team must verify that the HACCP system is working the way that it should. This is called verification. Various methods, procedures, or tests are used to validate that the HACCP system is operating as intended.

Verification may include:

- Observing monitoring procedures,
- Calibrating monitoring equipment (e.g., thermometers, pH meter, x-ray machine),
- Sampling product for microbiological testing, and
- Reviewing HACCP records.

HACCP systems are rarely perfect, and verification allows the company to identify the flaws in the process and fine-tune them to correct the system.

Prerequisite programs

Even though HACCP is highly effective, it is not a stand-alone food safety system. It is just one building block of a complete food safety program. Good sanitation, personal hygiene, and pest control programs are all necessary to safe food production. These are sometimes referred to as prerequisite programs because they need to be in place before a HACCP system can be effectively implemented.

In conclusion

As you can see by the information in this chapter, HACCP is a complicated process. When systematically broken down into many steps, however, it is a practical and manageable tool for food safety. Food producers across the United States use HACCP to ensure the safe management and production of food, so consumers of the product, people just like you and me, can eat safe food every day.

Name ____

Date ____

HACCP REVIEW

1. HACCP stands for:
   a. Hazard Analysis and Critical Control Points
   b. Hazards and Allergens Cause Certain Problems
   c. Having any Affects Causes Control Policies
   d. None of the above

2. The purpose of HACCP is to:
   a. Make you do more work
   b. Eliminate potential hazards from food to make it safe to eat
   c. Clean equipment properly
   d. None of the above

3. HACCP identifies which types of hazards:
   a. Biological
   b. Chemical
   c. Physical
   d. All of the above

4. Which U.S. government agencies require HACCP:
   a. United States Department of Agriculture
   b. Food and Drug Administration
   c. Both (a) and (b)
   d. Neither (a) nor (b)

5. How many HACCP principles are there:
   a. Five
   b. Three
   c. Seven
   d. Eight

6. Which is not a principle of HACCP:
   a. Hazard analysis
   b. HACCP team
   c. Monitoring
   d. Recordkeeping

7. Which is an example of a physical hazard:
   a. Glass
   b. Cleaning chemicals
   c. Bacteria
   d. None of the above

8. Which type of hazard is the cause of foodborne illness:
   a. Physical
   b. Biological
   c. Chemical
   d. None of the above

9. A critical control point (CCP) is:
   a. Identifying hazards and preventive measures
   b. A point, step, or procedure in a food process at which a hazard can be controlled
   c. Product sampling and testing
   d. All of the above

10. HACCP records must be:
   a. Accurate
   b. Signed
   c. Dated
   d. All of the above

SANITATION

All food companies are required to maintain sanitary conditions throughout the plant. The reason for this is clear. You can't make 'clean' food in a dirty plant, and therefore; good sanitation ensures the safe production of food.

Sanitation refers to *all* of the practices and procedures used to keep the plant clean and products free of contaminants and potentially harmful bacteria. This includes following good personal hygiene practices, cleaning equipment and work areas, maintaining clean work areas, implementing a pest control program, etc. Understanding your role and responsibilities regarding each of these sanitary operations will help keep your company clean at all times.

Sanitation basics

One of the most common images associated with sanitation is cleaning and sanitizing equipment and work areas. The reason for cleaning and sanitizing is to prevent the growth of harmful bacteria that can contaminate food and cause foodborne illness. Generally, a basic three-step process is used during sanitation operations.

Rough clean

The first, and perhaps most important part of sanitation, is the rough clean. This step involves removing the visible food residue from equipment and work areas. If equipment is not cleaned properly, remaining bacteria will continue to grow and contaminate food that comes into contact with the unclean equipment.

All food particles should be placed into an appropriate container for disposal.

It is often necessary to disassemble some equipment, like mixers and saws, to get at enclosed parts and hard-to-reach areas for proper cleaning.

Next, any food particles and juices remaining on equipment must then be rinsed off with water so that the cleaner and sanitizer can effectively do their job.

Clean

The second step is to clean with a detergent. The detergent, or other cleaning chemical(s), can be applied in different ways. They can be:

- Mixed with water in a bucket for cleaning by hand with a brush or cloth,
- Metered into a system that can be used through a hose,
- Applied as a foam, and
- Used in a clean-in-place system where chemicals are circulated through an enclosed system to clean the inside of tanks, lines, or other parts of equipment.

When cleaning equipment, small parts should be placed into a tub with detergent to soak. Small parts and hard-to-reach areas of equipment need special attention, therefore; they need to be thoroughly scrubbed.

Meanwhile, large pieces of equipment should be soaked with cleaner. The cleaner must be left on the equipment for a specified amount of time for it to work properly. After cleaning, all equipment and parts should be thoroughly rinsed with water and inspected for cleanliness.

Sanitize

The third and final step is to sanitize with chemicals designed to kill bacteria. After the sanitizer has been applied to the surface of equipment, it may or may not be necessary to rinse it off with clean water. Equipment should not be reassembled until it is needed for use.

It's extremely important not to rush through any of these steps, even if you think equipment looks clean to you. Remember, you can't see bacteria without using a microscope. You must make enough time to keep products safe from contamination.

If you are required to perform any of these sanitation steps, follow the procedures you are trained to use for each piece of equipment and/or work area.

Chemicals

A wide variety of chemicals are used in food companies to keep them clean. These chemicals are both safe and necessary in the sanitary production of food. However, they can be a contamination hazard if not used, handled, and stored properly.

Use

Chemicals must always be used according to the manufacturers recommended guidelines or label directions and company sanitation procedures. Also, don't forget to use proper personal protective equipment (PPE) when cleaning and sanitizing. Wear gloves, eye protection, and other PPE that may be necessary while using each chemical.

While working with or around chemicals in your workplace, you should:

- Put away or cover food and packaging materials before sanitation operations begin,

- Use only the chemicals approved by your company and specified for each task,
- Only use chemicals for which you've been properly trained on their specific usage requirements, and
- Always make sure proper labels are affixed to containers to prevent mix-ups and accidental contamination of product.

Also, know the location of the material safety data sheets for the chemicals that you work with.

Storage

Cleaning and sanitizing chemicals must be stored safely away from food production and storage areas to prevent contamination. Keep chemical storage areas properly secured at all times. A careless mistake could cost somebody their life!

Equipment

The use and storage of cleaning and sanitizing equipment is another important sanitation activity.

Cleaning equipment like mops, brushes, and rags must be cleaned before being put away. Residue left on these items will provide a source of food for bacteria to grow. As a result, the next time that mop, brush, or rag is used, bacteria will be spread onto equipment, rather than be removed from it, as cleaning and sanitizing activities are intended to do.

Cleaning equipment should be cleaned and hung to dry after each use to eliminate the damp and wet conditions bacteria thrive in. Some companies use disposable equipment (e.g., rags or cloths) to avoid possible contamination during sanitation activities.

Some companies have cleaning equipment just for certain areas of the plant. This helps prevent cross contamination of work areas. For example, certain buckets, mops, and brushes may be used only in raw product areas, while a different set is used only in finished product areas. If this is the case, the equipment may be color-coded or labeled accordingly.

Only use the proper set of equipment for each work area you're cleaning and store it in its designated place.

Housekeeping

A sanitation crew is likely responsible for cleaning equipment and work areas on a daily basis, but nearly all employees are involved in cleaning and sanitizing activities in some way.

One way is to keep your work area clean. By doing so, you maintain sanitary conditions in the plant. This helps ensure the safe production of food.

You should follow these simple housekeeping tasks to keep areas clean and tidy at all times:

- Keep floors clean by removing garbage and product waste promptly;
- Clean up spills immediately;
- Keep items such as packaging materials and boxes in designated storage areas to decrease the chance of contamination;
- Put items such as hoses, brooms, and carts away after each use; and
- Keep work areas as dry as possible to prevent bacteria from growing.

These simple housekeeping tasks will help keep your company clean and the food you produce safe to eat.

Garbage

Proper storage and disposal of garbage is necessary to minimize the development of odors, minimize infestations of insects and rodents, and prevent the soiling of production areas. Improperly handled garbage creates nuisance conditions, makes housekeeping difficult, and may be a possible source of contamination of product and equipment.

To ensure proper storage and removal of garbage, follow these simple guidelines:

- Check garbage containers for cracks or holes to make sure they are always kept in good repair.
- Line containers with plastic, if necessary, to prevent leakage.
- Clean and sanitize garbage containers as often as necessary to prevent infestations of insects or rodents.
- Keep your work area clean at all times.
- Remove garbage as soon as possible.
- Avoid carrying garbage through other food production areas.
- Always wash your hands after handling or removing garbage.

SSOPs

Standard operating procedures (SOPs) are established procedures for conducting a process. SOPs are usually in writing and specify the method, duration, and frequency of the given process the SOP dictates. SOPs should cover the who, what, when, why, and how for each procedure. Every company must develop its own unique set of SOPs for given processes because SOPs are specific to the activity and the environment in which they are performed.

There are SOPs for many activities in food manufacturing. Sanitation SOPs (SSOPs) describe the cleaning and sanitizing methods conducted in the plant to prevent the contamination of product. SSOPs are part of every company's food safety system.

SSOPs must be written to include the following:

1. A description of each daily sanitation procedure conducted before, during, and after operations.
2. Sanitation procedures conducted before the start of operations each day that address the cleaning of all food contact surfaces, equipment, and utensils.
3. How often each sanitation procedure will be done.
4. Identification of the individuals responsible for implementing and monitoring daily sanitation activities.
5. Records maintained on a daily basis to document the implementation and monitoring of SSOPs.

These sanitation procedures must be monitored on a daily basis. You may be required to participate in the implementation or evaluation of your company's sanitation program. For example, you may clean, inspect, teardown, or setup processing equipment.

You may also be required to fill out SSOP records. If so, be sure they are filled in as required and are accurate, legible, signed, and dated accordingly. These records are proof that your company is manufacturing safe and wholesome product.

In conclusion

Sanitary food production presents challenges to any company. There are many procedures used to keep the food production environment clean. As long as you follow sanitation procedures and are aware of the sanitation issues you work with, you can contribute to the safe production of food.

Name ____

Date ____

SANITATION REVIEW

1. Sanitation refers to:
   a. Only cleaning of equipment and food work areas
   b. Only sanitizing of equipment and food work areas
   c. All of the practices and procedures used to keep the facility clean and the food produced uncontaminated
   d. Cleaning and sanitizing of non-food work areas

2. Which of the topics are related to sanitation:
   a. Personal hygiene and cleanliness
   b. Equipment and work area clothing
   c. Pest control
   d. All of the above

3. Which does not require strict sanitation procedures be followed:
   a. Food production equipment
   b. Employees' automobiles
   c. Food production work areas
   d. Employees' personal hygiene

4. Food residue left on equipment is:
   a. A source of food for bacteria
   b. A source of chemical contamination
   c. Ok, it does not need to be cleaned off
   d. All of the above

5. How many steps are in a basic cleaning and sanitizing procedure:
   a. One
   b. Two
   c. Three
   d. Four

6. The basic steps of cleaning and sanitizing include:
   a. A rough clean
   b. Cleaning with a chemical
   c. Sanitizing
   d. All of the above

7. Methods of applying cleaning and sanitizing chemicals include:
    a. Mixing with water and apply with a brush
    b. Metering and apply through a hose
    c. Applying as a foam
    d. All of the above

8. Cleaning and sanitizing equipment and chemicals should be:
    a. Left any old place in the plant after use
    b. Used, handled, and stored properly
    c. Put away when still wet
    d. None of the above

9. When storing cleaning and sanitizing equipment, make sure to:
    a. Clean the equipment after each use
    b. Dry the equipment thoroughly before putting it away
    c. Store the equipment only in its designated storage area
    d. All of the above

10. When using cleaning and sanitizing chemicals always:
    a. Read and following the label instructions carefully
    b. Do whatever way you want
    c. Use as much or as little as you want
    d. None of the above

TIME & TEMPERATURE CONTROLS

Bacteria exist everywhere in nature. They are found in soil, air, water, and the foods we eat. When bacteria have a source of food, moisture, time, and favorable temperature conditions, they grow rapidly to levels that can cause serious foodborne illnesses. Therefore, understanding the important role time and temperature plays in keeping food safe is critical to your job.

The "Danger Zone"

Bacteria grow most rapidly between 40 and 140 degrees Fahrenheit (°F). This temperature range is often called the "Danger Zone." Food left too long in the temperature "Danger Zone" can make a person very sick and even kill them.

This is precisely the reason why your company has time and temperature processes in place to control the growth of harmful bacteria like *Listeria*, *E. coli*, and *Salmonella*, and to slow the spoilage of product. There are many different types of time and temperature controls and each is used for different reasons on different types of products. Several common time and temperature controls are described below.

Common time and temperature controls

Which time and temperature process used in making a product depends on the type of food being processed or produced, what packaging is to be used, how it will be sold, and many other factors.

Cooking

Cooking is the most common time and temperature control that is used in food processing. Products that are properly cooked are heated for a long enough time and at a high enough temperature to kill harmful bacteria. The only way to accurately tell if a product has reached a safe internal temperature is to use a thermometer.

Common cooking practices include:

- Baking
- Boiling
- Frying
- Steaming
- Broiling
- Roasting

Freezing

Freezing involves cooling food to a temperature below 28 °F. Freezing keeps food safe by stopping the growth of bacteria, but it does not kill it. Once thawed, bacteria can again become active, multiplying under the right conditions to levels that can cause foodborne illness. That's why specific procedures must be followed when thawing frozen food to make sure it remains at a safe, constant temperature.

Pasteurization

Pasteurization involves heating certain foods and liquids to a temperature between 131 and 158 °F, holding it at that temperature, and then cooling it again. This process partially sterilizes products by effectively destroying bacteria. However, pasteurized foods and beverages must still be stored under refrigeration. Pasteurization is used to process milk, cheese, wine, beer, and orange juice.

Refrigeration

Refrigeration cools food to a temperature cold enough to slow the growth of bacteria. This is why many food companies are cooled to a temperature around 40 °F. Refrigeration keeps processing rooms, coolers, and holding areas at a temperature adequate to keep food safe.

Blanching

Blanching is a brief, partial cooking to prevent deterioration of food. Blanching is accomplished by exposing food to a heat treatment to inactivate the naturally occurring enzymes found in food that can lead to spoilage. This process is often used prior to freezing foods like vegetables, such as peas and corn.

Ultra High Temperature (UHT)

UHT is a process in which food is heated in complicated equipment to a very high temperature for a very short time. The product is then packaged in a sterile environment. The finished product is considered shelf stable and does not need to be refrigerated.

Thermal processing

Although all heat processes are thermal processes, the term "thermal processing" is given to specific processes like canning and bottling. This process involves filled containers being heated to a specific temperature for a specific length of time. The finished product is considered sterile and can be stored without refrigeration. Just a few examples of these types of foods include canned vegetables and fruits, soups, and pouched juices.

Irradiation

During irradiation, foods are exposed briefly to a radiant energy source such as gamma rays or electron beams. Irradiation is not a temperature control, but the process can kill harmful bacteria and greatly reduce the risk of foodborne illness. Irradiation is approved for use on foods such as fresh fruits and vegetables, spices, and meat and poultry.

All time and temperature controls must always be followed according to established company procedures to adequately protect product.

Thermometers

Using a thermometer is a reliable way to make sure product has been properly processed to destroy harmful bacteria, however; to be effective thermometers must be used and calibrated correctly.

Many employees use thermometers to measure the temperature of products on a daily basis. If a thermometer is inserted incorrectly, or placed in the wrong area, the reading may not accurately reflect the internal temperature of the product. The following are key elements you should follow for accurately measuring product temperature:

- Take the temperature of several items in various places for irregular shaped products or those that vary in size like hams or roasts.
- Stir product, if necessary, before taking the temperature.
- Place the stem or probe of the thermometer in the thickest part of the product.
- Do not rest the stem or probe on a bone or the side of a container because this may give an inaccurate reading.
- Make sure the entire sensing area on the device is completely submerged in the product, if possible.

Also, properly wash and sanitize thermometers prior to use and in between uses and use a newly calibrated thermometer as often as required.

If you're required to take product temperatures during your workday, be sure you follow correct company procedures. This is the only way to gauge whether product is sufficiently cooked, and therefore, safe to eat. The people who eat the food you make are depending on you!

Daily work practices

As an employee you are the closest to production. You likely monitor or maintain time and temperatures processes of products. Follow these simple rules to make sure correct time and temperature controls are met for the safety of products:

- Always follow established work procedures and practices for the product you are processing.
- Keep doors to coolers, freezers, holding areas, warehouses, processing rooms, and other cooled work areas closed. This will help ensure that a constant temperature is maintained.
- Never alter the time it takes to heat, blanch, cook, or cool a product.
- If in doubt as to whether a food had reached the correct processing temperature, use a thermometer to check.
- When timing of procedures or practices is necessary, pay careful attention to the clock to be sure to meet the correct time requirement.
- If you are responsible for the time and temperature controls at a CCP under a HACCP system, carefully follow all required procedures.

If you think a time or temperature control may have been violated, tell your supervisor or quality assurance/control person. He or she can test the product or make a decision as to what to do.

In conclusion

Because you know how time and temperature controls affect the growth of bacteria in food, you can protect yourself and customers from foodborne illness by properly handling, heating, and storing foods at safe temperatures.

Name ____

Date ____

TIME & TEMPERATURE CONTROLS REVIEW

1. The temperature "Danger Zone" is between:
   a. 20 and 120 °F
   b. 30 and 130 °F
   c. 40 and 140 °F
   d. 50 and 150 °F

2. Time and temperature controls are used to properly:
   a. Cook food
   b. Cool food
   c. Store food
   d. All of the above

3. Time and temperature controls include:
   a. Cooking
   b. Freezing
   c. Pasteurization
   d. All of the above

4. The purpose of time and temperature controls is to:
   a. Produce food faster
   b. Control the growth of potentially harmful bacteria
   c. Create a warm, moist environment
   d. Make food taste good

5. Which bacteria can be controlled by time and temperature:
   a. *Listeria*
   b. *E. coli*
   c. *Salmonella*
   d. All of the above

6. Which is the most commonly used time and temperature control in food processing:
   a. Cooking
   b. Irradiation
   c. Thermal processing
   d. All of the above

7. Cooling food to about 40 degrees Fahrenheit is called:
    a. Freezing
    b. Pasteurizing
    c. Refrigerating
    d. All of the above

8. The first rule to follow for time and temperature controls is:
    a. Being close to target is good enough
    b. Always follow your company's established procedures and practices for the food you are making
    c. If you make a small mistake, it won't matter much
    d. None of the above

9. To make sure correct time and temperature controls are met:
    a. Keep doors closed to coolers, freezer, holding areas, warehouses, processing rooms, any other cooled work areas
    b. Never alter the specified time to heat, blanch, cook, or cool a product
    c. Use a thermometer to check if a food has reached its correct processing temperature
    d. All of the above

10. If you think a time or temperature control may have been violated:
    a. Tell your supervisor or quality control person, so they can test the batch or make a decision as to what to do
    b. Throw the batch away when no one is looking
    c. Don't tell anyone
    d. None of the above

FOREIGN MATERIAL

Foreign material detection is a way to locate physical objects that might be in product, but that should not be there.

It is a process that must be performed on all products at certain stages during the production process. This helps ensure that products don't contain potentially harmful foreign objects. Consumers want food that won't cause them injury.

You may be familiar with some foreign material detection devices or methods. In fact, you very well may be the person who performs inspections or operates special equipment. If so, it's very important that you follow all company procedures.

What is foreign material?

There are a wide variety of foreign materials. They include items that may come into the plant in raw materials, may be found in the plant, or may be introduced by employees. The table below lists some common foreign materials and their sources.

Foreign material	Source
Metal	• Nuts, bolts, and screws • Grinders • Mixers • Knives • Shovels • Meat hooks • Thermometers
Glass	• Light bulbs • Watch crystals • Bottles
Wood	• Crates • Pallets • Sawdust/ wood chips for smokehouses
Pests	• Environment • Incoming ingredients
Plastic	• Box/combo bands • Packaging materials • Broom handles
Hair, gum, jewelry, pens, cigarette butts, etc.	• Poor personal hygiene practices by employees

None of these foreign materials are supposed to be in food.

How to detect foreign material

Using in-line equipment for detecting and removing foreign material is a common method of inspecting and handling incoming raw materials. Similar in-line equipment, such as metal detectors, magnets, and occasionally x-rays, may be used during the production process or on the finished product before it is shipped. If equipment such as this is used, the operator must be trained on its proper use, and the company must do proper installation, regular maintenance, and calibration.

Another inspection method of incoming and outgoing product, perhaps the most common, is visual inspection. That means an employee, like you, looks at product to spot foreign material with the naked eye. This can be a highly effective technique if used properly.

Examples of visual inspections include:

- Inspecting tables or belts on a line to exam raw fruits and vegetables for stones, stems, leaves, wood, etc. before they are processed.
- Inspecting beef, chicken, or pork at receiving for cardboard, wood, dirt, and metal objects such as meat hooks, shovels, and thermometers.

Finally, there are several pieces of equipment designed for specific foreign material detection or removal functions as follows:

- **Magnet** — removes hazardous metals.
- **Metal detector** — detects various sizes of metal pieces.
- **X-ray machine** — visually detects any material in the product that appears abnormal.
- **Screen, filter, or sifter** — removes foreign objects larger than the size of the openings.
- **Aspirator** — removes material lighter than the product.
- **Riffle board** — removes stones from dry beans and field peas.
- **Bone separator** — removes bone chips from meat and poultry products.
- **Inversion equipment** — mechanically inverts bottles or cans so anything inside falls out.

It is likely that at least one of these pieces of equipment is something you use on the job.

When to conduct foreign material detection

It is important to ensure the safety of products at all times. The most important times to conduct foreign material detection is when stock is coming into the plant and when finished product is about to be packaged or is about to leave the plant. However, product must be inspected at all points during production, not just at the beginning and at the end.

Incoming raw materials

Careful attention must be paid to incoming raw materials. They can harbor pests, dirt, and debris. The pallets and crates that they are carried on or in are a frequent source of wood splinters, nails, and other hazards.

The following is a checklist to help you inspect incoming raw materials to make sure suppliers have delivered a safe and wholesome product:

- Inspect the delivery truck for cleanliness. Unclean trucks could present unsanitary conditions that could lead to food contamination.
- Check the rear of the trailer for splash protection. This guards against the possibility of highway pollutants seeping into openings and contaminating product.

- Examine the truck for physical defects. Cracks, crevices, and doors that do not tightly seal could leak air and alter the internal temperature of the trailer, causing product to spoil.

- Check product temperatures to make sure refrigerated foods are 40 °F or below and frozen shipments are 0 °F or below. In particular, check product near the back of the truck, temperatures are often warmer in this section.

- Inspect the trailer for signs of temperature abuse. Pools of ice and ice crystals may suggest that the shipment may have been exposed to warm temperatures during shipping.

- All food should be wrapped or in a closed container to prevent contamination by dust, dirt, or pests.

- All packages or shipping containers, including pallets and crates, should be clean and free of tears, crushed sides, or wood splinters.

- Products should not appear dirty, have an unusual color, or emit strong odors. For example, fresh meat should be pink and have a firm texture, not sticky or slimy. If you discover product you believe to be suspect, contact your supervisor immediately.

- Check the expiration and cut dates on items like dairy products and fresh meat and poultry.

- Examine the trailer for pesticides, cleaning supplies, and other chemical contaminants that must be transported separately.

- Review any shipping instructions that might indicate special care or caution be given to the shipped product, such as requiring that all the containers be shrink-wrapped.

- Examine the bill of lading to make sure the correct items have been received.

By following these simple inspection tips, you can make sure the product your company makes is safe from beginning to end.

Ongoing detection

Ongoing foreign material inspection and detection is also important. There are many points during production when foreign materials could get into product. You can take steps to protect product from foreign material each day by:

- Always practicing good personal hygiene. This keeps items like jewelry, pens, and gum from getting into food.
- Carefully add ingredients to batches of product. For example, when opening and dumping bags, be careful to keep pieces of paper or plastic from getting mixed in with the product.
- Maintain a constant awareness of your work area and watch for signs of potential contamination sources like flaking paint, rusty equipment, and loose screws, nuts, or bolts.
- Practice good housekeeping at all times to keep work areas clean and free of debris.

Outgoing food

Food on its way out of the plant is the last chance to detect foreign material before the food leaves the plant and is distributed for consumption. In addition, since all stages of the production process are complete by this point, outgoing food inspection is done after the last opportunity for foreign materials to contaminate the food.

In conclusion

When foreign material detection devices are used, hazardous objects will be found. But just because these controls are in place, doesn't mean that you aren't an essential part of the process. You must also do your part to detect foreign material to make food that's safe to eat.

Name ____

Date ____

FOREIGN MATERIAL DETECTION REVIEW

1. Foreign material detection is:
   a. Finding objects in your morning coffee before going to work
   b. Locating objects in food that should not be there
   c. Sanitizing of equipment and food work areas
   d. None of the above

2. Foreign material detection is important because:
   a. It helps ensure that products don't contain foreign objects
   b. It is job security for employees
   c. It helps detect bacteria
   d. None of the above

3. Foreign materials can come from:
   a. Raw materials
   b. Inside the plant
   c. Employees
   d. All of the above

4. Types of foreign materials include:
   a. Metal
   b. Glass
   c. Wood
   d. All of the above

5. Poor personal hygiene practices are the cause of which types of contaminants:
   a. Chewing gum
   b. Hair
   c. Jewelry
   d. All of the above

6. It is important to conduct foreign material detection:
   a. When raw materials are received and when food is about to leave the plant
   b. During the cooking process
   c. After the food has arrived at the store
   d. All of the above

7. Incoming raw materials can contain which foreign materials:
   a. Pests
   b. Dirt
   c. Debris
   d. All of the above

8. Which equipment is used to detect foreign materials:
   a. A thermometer
   b. Magnets and metal detectors
   c. A refrigerator
   d. None of the above

9. Visual inspection means:
   a. Looking through a magnifying glass
   b. Using magnets and metal detectors
   c. Looking for foreign materials with the naked eye
   d. None of the above

10. Which control measures help prevent foreign materials from getting into food:
    a. Good personal hygiene practices
    b. Inspection of raw materials before adding them to food
    c. Good housekeeping measures
    d. All of the above

CROSS CONTAMINATION

Cross contamination is the transfer of harmful contaminants to product. This contamination can come from employees, equipment, utensils, other products, etc. Preventing cross contamination helps keep food safe to eat.

Contaminants are items in a product that should not be there. This includes wood, glass, metal, sanitation chemicals, allergenic foods like peanuts, and harmful bacteria like *E. coli*. These types of hazards are discussed in more detail so you can help prevent cross contamination.

Chemicals

Chemical are a particularly hazardous type of food contaminant.

A wide variety of chemicals are routinely used in food production facilities to keep them clean. Many of them, if used improperly, can be poisonous to humans. This includes:

- Pesticides
- Lubricants
- Paints
- Cleaners
- Sanitizers

Many of these chemicals must be used in close proximity to or in production areas. At the same time, they must be kept from contaminating food.

The following are guidelines when working with or around chemicals in your workplace to prevent cross contamination:

- Don't use chemicals if you have not been properly trained.
- Always put food and packaging materials away or make sure they are properly covered before cleaning or applying chemicals.
- Use only approved chemicals for the task you are doing.
- Use each chemical according to the manufacturers recommended guidelines or instructions printed on the label.

- Store chemicals in their designated spot when not in use. There they will not spill, leak, or be accidentally added to the product being made.
- Always make sure a proper label is affixed to the container.

- Make sure production areas are cleaned, rinsed, and free of chemical residue after sanitation, maintenance, or pesticide application.
- Know where the material safety data sheets are kept in your workplace.

When in doubt, talk to a supervisor for more information on the chemicals you handle at work.

Bacteria

Bacteria do not have legs, so they cannot move around on their own. They have to hitch a ride on people, equipment, packaging materials, etc. to move from one place to another. Therefore, obeying personal hygiene rules, like handwashing and using footbaths and hand dips, are extremely important in preventing bacterial cross contamination.

Another good control to help prevent this type of cross contamination is to separate raw product from cooked product. A raw product can contain numerous types of bacteria. If the product is cooked properly, bacteria are killed. Therefore, cooked product should never come in contact with raw product because it may not be cooked again before it is eaten.

There are usually separate raw and finished work areas and/or separate equipment in raw and finished product areas in the plant. The equipment may even be color-coded to help you identify which equipment should be used in which area. This prevents accidental contamination.

Physical hazards

Physical contaminants can be just as dangerous as chemicals and bacteria, and therefore; you should follow these work practices:

- Be sure that a thorough clean-up is done after maintenance work on a piece of equipment is complete to ensure that nuts, bolts, or metal filings can't contaminate product.
- Store personal effects, like purses, cigarettes, and jewelry, separate from production areas. When stored separately, these items have no opportunity to contaminate food.
- Keep pests out of the plant.
- Inspect incoming raw materials to detect any physical hazards (e.g., wood, metal, glass) before the ingredient is used.

It's important to keep as non-food and non-work-related items separate from food production areas, to prevent contamination of ingredients and finished product.

Horseplay

Horseplay is a serious issue that can jeopardize your safety and that of your co-workers. However, it can also jeopardize the safety of the products you make. The connection may not seem obvious, but incidents of horseplay can actually contaminate food.

This cross contamination can occur by:

- **Playing "basketball" with garbage cans.** Items such as paper towels and gloves can bank off the edges of a receptacle and fall into product containers and onto production lines.
- **Using hoses to squirt one another.** Spray from non food-contact surfaces, like the floor, can contaminate product, production lines, and other equipment.
- **Joy riding on racks of products.** Contact with dirty shoes, clothing, and skin would directly contaminate product.
- **Driving forklifts and similar vehicles recklessly.** Open racks of products can easily bump or swing up against walls and contaminate product during transport throughout the plant, and boxes can be punctured by pieces of wood from pallets if handled carelessly.
- **Throwing food in production areas.** Airborne food can easily miss its "mark" and sweep across contaminated surfaces, such as hardhats, before falling onto production lines or into containers of edible product.

Obviously, these types of horseplay activities are not only dangerous, but can seriously compromise the safe production of food.

Detection and elimination

Cross contamination can occur when rules are disobeyed, control measures are inadequate, or through human error or accident. That's why we cannot rely solely on the prevention of cross contamination. We must also have practices in place to check food during and after production to detect cross contamination that may have occurred.

Some of these detection practices include:

- Visual inspections of product;
- Use of metal detectors, magnets, x-rays, or other detection equipment;
- Use of screens, filters, or sifters; and
- Microbiological sampling and testing.

In conclusion

Preventing cross contamination means following the food safety rules at your company to prevent cross contamination of food. But the simplest rule to follow is this: Separate — don't cross contaminate!

Name ____

Date ____

CROSS CONTAMINATION REVIEW

1. Cross contamination is:
   a. Contamination of food with crosses
   b. Contamination of food with other materials
   c. Contamination of food with ingredients
   d. None of the above

2. Which is not a possible contaminant:
   a. Pesticides
   b. Wood
   c. Bacteria
   d. Moon beams

3. Chemical contaminants include:
   a. Water
   b. Cleaning chemicals
   c. Milk
   d. All of the above

4. To prevent chemical cross contamination always:
   a. Store chemicals in their designated spot when not in use
   b. Make sure food production areas are cleaned, rinsed, and free of chemical residue after sanitation
   c. Put food and packaging materials away or cover them before cleaning or applying pesticides
   d. All of the above

5. Bacteria can contaminate food by hitching a ride on:
   a. People
   b. Equipment
   c. Packaging materials
   d. All of the above

6. To prevent bacterial cross contamination always:
   a. Wash your hands
   b. Use footbaths and hand dip stations
   c. Separate raw and cooked products
   d. All of the above

7. Physical hazards include:
   a. Metal
   b. Dust
   c. People
   d. Water

8. Horseplay includes:
   a. Driving a forklift recklessly
   b. Using hoses to squirt co-workers
   c. Throwing food
   d. All of the above

9. Which is not used to detect cross contamination in food:
   a. Metal detector
   b. Screwdriver
   c. Visual inspection
   d. Screens and sifters

10. Cross contamination occurs when:
   a. Rules are disobeyed
   b. Control measures are inadequate
   c. There is human error or accident
   d. All of the above

ALLERGENS

People can eat most food safely. But, imagine what it would be like if eating a peanut butter sandwich or drinking a glass of milk left you vomiting, gasping for breath, and furiously scratching a fresh crop of hives. For people with food allergies, that's reality.

Deadly consequences

A food allergic reaction is caused when the body's immune system responds to a foreign protein found in the offending food. Typically, the allergic reaction has a very rapid onset. Food allergy sufferers commonly experience:

- Vomiting,
- Diarrhea,
- Hives or rashes,
- Stomach cramps,
- Difficulty breathing,
- Tingling sensation in the mouth,
- Swelling of the tongue and throat, and

- Loss of consciousness.

The most severe reactions may even result in death.

Common allergens

Many different foods can cause an allergic reaction, however; the majority of food allergies are caused by:

- Peanuts,
- Soybeans,
- Milk,
- Eggs,
- Fish,
- Wheat,
- Shellfish (e.g., shrimp, crab, lobster), and
- Tree nuts (e.g., almonds, walnuts, pecans, cashews).

For sensitive individuals, the presence of allergens in food is potentially life-threatening. Currently, there is no cure for food allergies. The only way to treat food allergies is to avoid foods that trigger reactions.

Prevention is the key

Most food companies are aware of how serious food allergies can be, and as a result, have made changes in their manufacturing and labeling practices. Unfortunately, food allergens can still contaminate products by improper:

- Use of ingredients,
- Scheduling of product runs,
- Use of rework, or
- Clean-up or sanitation of equipment.

Ingredients

To avoid accidental contamination, check shipments at receiving to determine that the correct ingredients and raw materials have been delivered as indicated on the order form or bill of lading.

Storing allergenic ingredients separately from other ingredients is also a way to prevent contamination of other products.

Also, always check the ingredients before they're added to a batch of product. Each ingredient should be checked against the formula or recipe sheet and marked off as they are added to the mix. Labels on ingredient bags and containers must be carefully read.

Processing

Follow all company food safety procedures when making products to avoid allergen contamination, which may include:

- Using separate or color-coded equipment (e.g., scoops, pails, totes, bins, hoppers, conveyors) only for products that contain a known allergen. This prevents residue that may be left on equipment from contaminating other products.
- Visually inspecting equipment for allergen residue or build-up from previous production runs before processing begins of another product.

- Processing allergen-containing products at the end of the day or week so that there is time to clean the equipment before the next production run.
- Manufacturing products that don't contain an allergen before products that do.
- Adding the allergen ingredient near the end of processing.

Labeling

People with food allergies must always read the label of all foods, every time they shop, therefore; it's very important to verify that the correct labels are used on products when they're packaged.

Rework

All rework must be clearly labeled. It can only be used in a product that is compatible with the product being made at the time. For example, rework that has egg in it, can only be used in a product that also contains egg and is declared on the label as such.

Also, waste product that may contain allergens must be disposed of properly.

Sanitation

Proper cleanup following the processing of products that contain allergens is essential. Equipment may need to be disassembled and cleaned by hand to make sure hard-to-reach areas are free of allergen residue, and visually inspected after cleaning. All of these steps should be properly documented.

Also, ingredient bins or containers should be cleaned between uses in storage or pre-weight areas to prevent mixing and cross contamination.

Finally, if you work with any type of allergenic food, you must follow good personal hygiene practices. Even minute traces of allergenic foods can be transferred from your clothes or hands to another product. Be sure to change your gloves, apron, smock, or other garments before beginning work on a product that does not contain an allergen.

In conclusion

It takes the awareness and contribution of every employee to avoid this type of serious contamination. It's very important that you follow all company food safety rules and procedures concerning allergens. Allergen control is a necessary part of manufacturing a safe product for consumers.

Name ____

Date ____

ALLERGENS REVIEW

1. A food allergic reaction is caused by:
   a. Eating too much food
   b. The body's response to a foreign protein
   c. A reaction to harmful bacteria in the food
   d. All of the above

2. Sufferers of food allergies commonly experience:
   a. Vomiting
   b. Hives
   c. Difficulty breathing
   d. All of the above

3. Common food allergens include:
   a. Peanuts
   b. Milk
   c. Fish
   d. All of the above

4. The only way to treat food allergies is to:
   a. Get a shot
   b. Lay down and rest
   c. Avoid foods that trigger reactions
   d. None of the above

5. Food allergens can contaminate products by improper:
   a. Use of ingredients
   b. Use of rework
   c. Clean-up or sanitation of equipment
   d. All of the above

6. To avoid accidental contamination of ingredients with allergens:
   a. Check shipments of ingredients and raw materials at receiving
   b. Store allergenic ingredients separately from other ingredients
   c. Always check the ingredients before they're added to a batch of food.
   d. All of the above

7. To avoid allergen contamination when processing products, you should:
   a. Visually inspect equipment for allergen residue or build-up from previous production runs before processing beings on another product
   b. Use separate or color-coded equipment for products that contain a known allergen
   c. Ad the allergen ingredient near the end of processing
   d. All of the above

8. Rework that contains an allergen must be:
   a. Clearly labeled
   b. Only used in a compatible product
   c. Both (a) and (b)
   d. Neither (a) nor (b)

9. Proper cleanup following the processing of products that contain allergens may include:
   a. Disassembling equipment
   b. Cleaning areas of equipment by hand
   c. Visually inspecting equipment after cleaning
   d. All of the above

10. When working with foods that contain allergens, you should always:
    a. Follow all company food safety rules
    b. Work as quickly as possible
    c. Avoid your co-workers
    d. All of the above

PEST CONTROL

There is a saying that the best defense is a good offense. This statement holds true when it comes to pest control at a food company. The best way to combat pests is to prevent them from getting into the plant in the first place.

And the reason why we don't want pests in a food plant is simple. Pests can carry and transmit disease and filth. Contaminated product can make people very sick.

Prevention is the key

To keep pests from getting in, the building should have no unprotected openings to the outside. Attractions like food and water around and in the facility should be eliminated as much as possible to prevent pests from finding a place of refuge or shelter.

However, despite the many precautions taken by the company or a professional pest control service, pests sometimes manage to get inside. In this case, there are several different methods of pest control that can be used. These include different kinds of insect, rodent, and bird control measures.

Insect control

An extremely common method of insect control is to use pesticides. Many of these pesticides are made of potentially harmful chemicals, so they must be handled, applied, and stored carefully. That's why many food companies employ professional pest control services.

If a professional service is not used, the person who applies pesticides at your company must be thoroughly trained in pesticide application and its hazards. Because of the potential danger of pesticides, its use is permitted only under precautions and restrictions that protect against the contamination of product, equipment, and packaging materials.

All product and ingredients must be put away before pesticides can be applied to an area of the plant.

There are three basic methods of pesticide application which you should be aware of in case it affects you or your work area, which includes:

- Fogging,
- Fumigating, and
- Crack-and-crevice spraying.

Other methods of insect control include the use of heat sterilization or insect electrocutors near entrances to the building.

Rodent control

Two main methods of controlling rodents in food plants are trapping and baiting. Trapping is using mouse, rat, or squirrel traps to capture and hold the rodent. Baiting uses traps with poison (rodenticide) that the rodent eats and carries away with it. Rodenticide poisons the rodent and causes it to infect the nest and die.

Traps generally have several practical advantages over baits, including the fact that they:

- Provide a record of the types of rodents trapped, what kind they are, how many there were, etc.;
- Identify problem areas in or around the plant;
- Pose no toxicity dangers like baiting does by using rodenticides; and
- Help determine the direction of travel and travel paths of rodents, thus making their elimination easier.

Baits are generally restricted to outside use because they contain dangerous chemicals. If bait is used to control rodents, generally:

- They can only be used in designated and authorized areas,
- They must be inspected and changed frequently, and
- Dead rodents must be disposed of properly to prevent the contamination of product.

Bird control

A less common type of pest is the bird. Sparrows, starlings, and pigeons are all considered potential pest birds around food companies. After eliminating potential food, water, and nesting areas, the best way to keep birds away is to keep windows, doors, and loading dock doors closed whenever possible.

Daily work practices

There are ways that you, as an employee, can help keep your company free of pests too.

Follow these work practices every day to support the pest control program in effect at your company:

1. Always maintain good housekeeping in all areas of the plant. Keep all work and storage areas clean, sanitary, and dry. Promptly clean up any spills that could attract pests.
2. Keep garbage areas clean both inside and outside the building. Make sure any leaks from garbage compactors are cleaned up.
3. Keep lids closed on trash receptacles and garbage barrels.
4. Inform pest control personnel or your supervisor whenever you see evidence of an infestation of any kind.
5. Keep food, snacks, beverages, and other belongings in a locker or designated storage area.

6. Never prop outside doors open for ventilation and always keep unscreened windows closed.
7. Closely inspect incoming raw materials for signs of pest infestation.
8. Watch for damaged traps or missing bait stations.
9. Tell a supervisor if you see dead insects or rodents in the plant so they can be disposed of properly.

In conclusion

We all know we don't want pests such as birds, rodents, cockroaches, flies, or other animals and insects around food. We understand that they pose a food contamination risk that only a proper pest control program can control. Follow pest control work practices to help keep your workplace free of pests and the products you make safe.

Name ____

Date ____

PEST CONTROL REVIEW

1. Pests include which of the following:
    a. Insects
    b. Birds
    c. Rodents
    d. All of the above

2. Pest control is part of your company's:
    a. Sanitation program
    b. Equipment maintenance
    c. Microbiological testing
    d. None of the above

3. Pests are attracted to:
    a. Food
    b. Water
    c. Garbage
    d. All of the above

4. The best way to combat pests is to:
    a. Combat them once they're inside the facility
    b. Keep them from getting into the food facility in the first place
    c. Detect them once they are in the food product
    d. None of the above

5. To keep pests from getting in:
    a. The facility should be tightly constructed
    b. Eliminate food and water sources
    c. Keep doors and windows shut
    d. All of the above

6. Insect control methods include:
    a. Insecticides
    b. Insect electrocutors ("zappers")
    c. Heat sterilization
    d. All of the above

7. The methods used to apply pesticides include:
   a. Fogging
   b. Fumigating
   c. Crack-and-crevice spraying
   d. All of the above

8. The main method of rodent control is:
   a. Baiting
   b. Trapping
   c. Both (a) and (b)
   d. Neither (a) nor (b)

9. A general rule you can follow to help control rodents through baiting include:
   a. Using them only in designated and authorized areas
   b. Inspecting and changing bait frequently
   c. Disposing of, or notifying the proper person to dispose of, any rodent bodies you discover
   d. All of the above

10. Which is not an example of good pest control:
   a. Cleaning up spills
   b. Leaving doors and windows open
   c. Cleaning up garbage
   d. Using pesticides

SECURITY

Food companies have to deal with safety risks such as theft, penetration of secure areas, assault, and vandalism. As a result, food companies have always had the need to incorporate a certain degree of security into their operations. But now, in today's environment, all food companies need to look at the issue of security in a different, and much more serious, light.

As the farm-to-table continuum continues to grow, so does the opportunity for intentional acts of contamination at vulnerable points in our food safety chain. The threat to our food supply can come from various sources. It may be from outside the company (e.g., terrorists) or from inside the company (e.g., disgruntled employees).

As a result, many food companies are stepping up measures to protect the safety of their property, products, and personnel. Therefore, it's in your best interest to follow your company's food safety and security procedures. The following information includes tips you should follow every day to help keep yourself and the products you make safe to eat.

Parking

Parking lots can be a very dangerous place because of unpredictable traffic patterns and careless drivers. But they can also be a prime target for thieves, carjackers, and vandals.

There's good news, however. There are safety precautions you can follow to protect yourself. When you enter an employee parking area you should park:

- Where you can clearly see around your vehicle.
- Where you can be seen (a highly visible location).
- In high-traffic areas of the lot, close to the building entrance, if possible.

Entering the building

Many workplaces now operate controlled access entrances. These are usually security checkpoints that are manned or unmanned. That means that there may be a:

- Security person or receptionist at the door screening employees entering the building, or
- Device that automatically scans employee identification (ID) cards or badges, and without a valid ID the door will not open.

If you have been issued an ID badge, make sure that you know where your security card is at all times. If your card is lost or stolen, immediately report this to your company's security office or other appropriate personnel like your supervisor or Human Resources department. Also, if you are asked to display your ID upon entering the building, cooperate with security by making sure they can easily identify you as the appropriate card holder.

Some companies require employees to use ID badges or cards to open the entrances of buildings. This is usually through the use of a scanning device. If this security measure is used by your company, it's important to make sure you only let yourself into the building.

If there is another person in back of you who wants to enter, insist that they use their own employee ID to scan and enter the building. Don't give building or gate access to strangers, including visitors and delivery persons. After all, you wouldn't give your home security code to a stranger, so don't do it at work.

Security cameras are often used to provide surveillance around building entrances. Don't look at the attempts by your company to protect you as an invasion of your privacy. Often the mere site of security equipment will deter criminals.

Security inside the building

There are certain areas in the plant which must maintain a certain level of security. These include warehouses holding raw materials and products, production areas, laboratories with hazardous chemicals and glassware, and storage areas for sanitation chemicals. An intentional act of sabotage could cause illness and even death to numerous people.

No matter where you work in the plant, you should:

- Monitor the equipment you work with (i.e., blenders, choppers, chill tanks, smokehouses, thermometers, ovens) to prevent sabotage or tampering of parts.
- Make sure that foreign material detection equipment (i.e., metal detectors, sifters, x-ray machines, magnets) is working properly at all times.
- Examine all ingredient containers, boxes, or bags for evidence that they have been opened or resealed.

- Make sure that product and rework is properly labeled at all times.
- Report missing stock (i.e., labels, packaging materials) to your supervisor.
- Keep areas where sanitation chemicals or other hazardous substances locked if they're supposed to be kept secure.
- Account for equipment (i.e., thermometers, knives, cutters, shovels) that you may use on a daily basis and report any missing items to your supervisor immediately.
- Report broken windows and locks.

Also, the personal items that you bring to work may not only pose a food safety risk, but a security risk as well. Do not bring lunch containers, purses, etc. onto the production floor which could be used to conceal chemicals, weapons, explosive devices, etc. Keep them in a designated storage area or locker.

Visitors

Any visitor (i.e., truck drivers, contractors, auditors, regulators, etc.) who enters the plant on a temporary basis is a potential security threat. For this reason, many food companies don't allow visitors in production areas unless it's absolutely necessary. If visitors are allowed in the plant, they usually must follow certain rules.

First, visitors should register at an entrance location before they move about the company. After verification that they are to be allowed in the plant, they may have to wear some form of identification. If so, the ID badge should be worn so that it is clearly visible at all times. Lastly, visitors should be escorted by an employee and not allowed to wander off alone.

After the meeting has taken place or the work is completed, visitors must return the ID badges and make a note in the visitor record that they have left the building.

Suspicious activities

Stay alert for and report suspicious activities while at work. Your personal security and that of the company is enhanced when you are aware of the routines in your workplace; the people, the equipment, the products, the processes, etc. That way, if there are any changes or abnormalities, they are easy to spot.

Watch for people or other employees:

- Attempting to access something that is normally denied them, such as maintenance shops, chemical storage rooms, or ingredient warehouses.

- Carrying a weapon (unless they are authorized to do so).
- Acting nervous, anxious, or secretive.
- Staying unusually late after the end of their shift.
- Arriving unusually early.
- Loitering outside the building.
- Bringing cameras to work.
- Making unusual requests, demands, or threats.

If you witness any of these activities, report what you see to a supervisor or other management personnel immediately.

In conclusion

Every employee has the responsibility to be aware and vigilant while at work. This vigilance is not paranoia. It is simply a way to protect yourself, fellow employees, product, and your company.

Name____
Date____

SECURITY REVIEW

1. Security risks at your company can include:
   a. Theft
   b. Assault
   c. Vandalism
   d. All of the above

2. Parking lots can be dangerous because of:
   a. Unpredictable traffic patterns
   b. Careless drivers
   c. Hidden areas that conceal criminals
   d. All of the above

3. When you enter a parking lot you should park:
   a. Where you can clearly see around your vehicle
   b. Where you can be seen
   c. Close to the entrance, if possible
   d. All of the above

4. If there is another person in back of you who wants to enter the building, you should:
   a. Let them in if you know them
   b. Insist that they use their own ID badge to scan and enter the building
   c. Check their ID badge to see if they are the card holder, and then let them in
   d. All of the above

5. An area of the plant which must be kept secure is a:
   a. Warehouse
   b. Production area
   c. Laboratory
   d. All of the above

6. To protect product while working, you should:
   a. Make sure metal detectors, magnets, etc. are working properly at all times
   b. Properly label product and rework
   c. Report missing stock
   d. All of the above

7. To protect yourself while working, you should:
    a. Report suspicious activities
    b. Report broken windows and locks
    c. Be aware of your surroundings at all times
    d. All of the above

8. Do not bring lunch containers or purses onto the production floor because:
    a. There is no room for them in your work area
    b. They could be used to conceal chemicals, weapons, and explosive devices
    c. Other employees might take your food
    d. None of the above

9. Suspicious activities include:
    a. Acting nervous, anxious, or secretive
    b. Taking pictures of work areas and equipment
    c. Threatening people
    d. All of the above

10. Visitors at your company should:
    a. Register before they move about your company
    b. Wear some form of identification
    c. Travel with the employee they have come to see
    d. All of the above

GLOSSARY OF FOOD SAFETY-RELATED TERMS

Allergen — A food that causes a severe allergic reaction.

Anti-bacterial — Directed at or effective against bacteria. Soap that is anti-bacterial eliminates or kills bacteria.

Bacteria — Minute microorganisms that are too small to be seen with the naked eye.

Biological hazard — Generally used to refer to the bacteria that pose a risk of causing foodborne illness.

Calibration — The adjustment of a testing machine or equipment to achieve accurate measurement.

Centers for Disease Control & Prevention (CDC) — A United States government agency that serves to protect the health and safety of people.

Chemical hazard — A hazard posed to food by chemicals in the food production or processing workplace.

Contagious — The condition of a disease such that it can be passed on from one thing to another.

Contaminant — Anything that can get into food that is not supposed to be there. Types of contaminants include glass, metal, wood, bone, hair, jewelry, bacteria, etc.

Cross contamination — The transfer of harmful contaminants to product.

Feces — Bodily waste discharged through the anus.

Food & Drug Administration (FDA) — A United States government regulatory agency that is part of the Public Health Service, under the U.S. Department of Health & Human Services. This agency regulates all foods, except meat, poultry, and egg products.

Foodborne illness — Illness associated with the consumption of contaminated food.

Food-contact surfaces — Surfaces that food comes into contact with. This includes utensils and surfaces of equipment.

Footbath — A container of water and/or germicidal solution provided to dip and clean feet/boots to ensure the maintenance of clean feet in food processing areas.

Foreign material — Physical objects that should not be in a product.

Gastroenteritis — Inflammation of the lining membrane of the stomach and the intestines.

Germinate — To sprout or develop.

HACCP — An acronym for Hazard Analysis and Critical Control Points.

HACCP plan — A logical, simple, effective, but highly structured system of food safety control.

Hand dip — A container of water and/or germicidal solution provided to dip and clean hands in food work areas to ensure the maintenance of clean hands while working with food.

Hair restraints — Anything used to pull back and/or capture hair to the head and protect it from escaping or falling into the food. Common types of hair restraints include hairnets, caps, beardnets, or other devices designed to restrain hair in the workplace.

Low-acid foods — Any foods, other than alcoholic beverages, with a finished pH greater than 4.6 and a water activity (a_w) greater than 0.85.

Microbes — Bacteria, molds, spores, and other microorganisms.

Microorganisms — Living things that are so small that they can only be seen with the help of a microscope.

Pathogens — Microorganisms (i.e., bacteria) that cause illness or death.